Esoteric Theories of Flora, Fauna & the Universe

CASH MATTOCK

ISBN: 0692815058
ISBN-13: 978-0692815052

CONTENTS

ACKNOWLEDGMENTS

Thank you very much Steve Bolton for your superior editing skills and to Jessica LK Photography for my wonderful portrait.

PROLOGUE

Loosen your grasp on everything you've been previously taught. You have been conditioned by this world to feel compartmentalized and unaware of the wonder that surrounds you. We are born into a world of ideas where everything we imagine is possible but as our comprehension grows, society is slowly blinding us. By the time we reach our full potential we can no longer see the world in this way and the magic is gone.

This book is about my theory of everything. My theory involves some concepts that stretch the limits of comprehension but I will present these ideas in a way that's less ambiguous and easier to visualize. I ultimately want everyone to be able to understand my theory and how it changes our perception of the universe. I believe this understanding is crucial. If everyone had a better understanding of their relationship with the universe then we could start to address the world's problems from a clearer perspective. Our bodies and our minds are products of our four-dimensional universe. We are not conscious of nor can see anything beyond this because everything about us has evolved in this environment, so we only have our imagination to fall back on. We also have to consider our limited ability to communicate these ideas so I plead with you to open your minds, because what I have to tell you isn't just about physics; it's also about the future of our species.

Claude Fayette Bragdon is an architect and author known for his design contributions in my native city of Rochester, New York. He also designed part of the Rochester Institute of Technology where I attended. His book titled "A Primer of Higher Space (The Fourth Dimension)" has been an inspiration. It was written during the time when the concept of extra dimensions was in its infancy. He used his knowledge of architecture and pure imagination to conceptualize extra dimensions. I feel that the age of quantum physics that we are in now is similar to the dimensional age of Bragdon's time. With all the recent discoveries credited to particle collisions and the popularization of the Higgs Boson I feel that the world is more open-minded to theoretical physics. It's finally time to share my theory.

I've tried to keep the explanations as simple as possible. You may notice that this book is intentionally devoid of mathematical formulas. I want the reader to appreciate the concepts - not the math. I will also review some basic physics principles to help bridge these concepts.

Many accomplished physicists agree that humans may never be capable of understanding everything - that there's a limit to what we can do. We need to throw our minds into the lost-and-found of the unperceived and see if we can get a little closer to the understanding of everything even though we may never get there.

PARTICLES, WAVES, SPIN, & THE STANDARD MODEL

Most people have seen the Periodic Table of Elements and understand that those elements are made up of atoms, which are made up of electrons, protons and neutrons. Protons and neutrons are at the center of the atom as the nucleus. The electrons orbit the nucleus in spherical shells, not in the way a planet orbits the sun. Electrons are where my story begins.

The electron was discovered in 1897 by Sir Joseph John Thomson. It is an elementary particle that has a negative charge. They are responsible for many things that we take for granted. Everything from electricity and magnetism to radio waves and X-rays all begin with the electron. Particles are the smallest bits that make up everything in the universe. We've had a very long time to study the electron compared to some of the other particles like quarks, which were discovered in 1968.

As electrons orbit the nucleus in these spherical shells their behavior is sometimes difficult for us to conceptualize. Electrons are inherently lazy in a sense, in that they want to occupy the closest available shell to the nucleus. If energy is introduced into the atom in the way of heat, for example, the electron jumps to a higher shell as the energy is absorbed and then jumps to the first lowest available shell as the energy is released. The energy released in this jump is released in the form of light. This is called a quantum leap (or jump). This makes sense because energy and mass are never lost or destroyed - but how do the electrons jump? The orbital wave path of an electron is very large in comparison to the size of the atom and the nucleus is small in comparison. As a result atoms consist of about 99.9999999999996% empty space. The only reason why matter is solid is because of the force holding it together. Actually, aside from the orbits in which the electrons traverse, the position and movement of electrons is not completely understood.

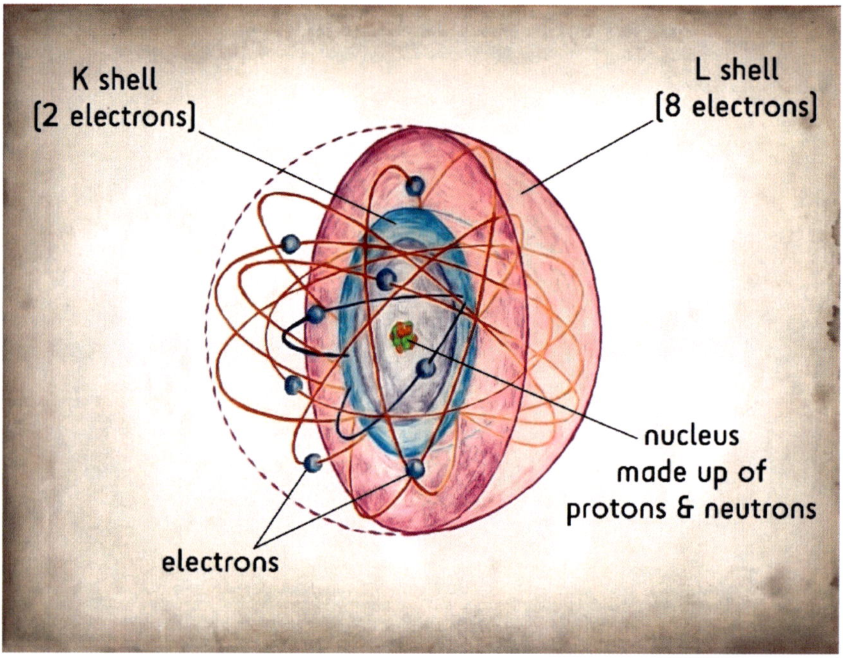

This is where things get a bit weird. One electron can occupy multiple spaces at the same time as a probability. This means that one electron can exist in two places at once until it's observed in one. This is called Superposition. It's also part of the Wave-Particle Duality. It's impossible to determine their position and movement at the same time. When you concentrate on a particle by slowing it down it's observed as a particle but you can't observe its movement. If you speed it up it's observed as a wave but you can't observe its location. It's very important to understand that particles are also waves, therefore so is matter. A baseball is like a very large particle. You can through it through the air and it moves in an arc or a straight line like a particle but it is also a wave with a very small wavelength. You and I are waves too but we are too massive to notice. The wavelength of us and the ball cannot be observed by the naked eye; it's a very small vibration.

Paul Dirac also believed it was possible that orbital shells existed in a negative space which would give the electrons even lower shells they could default to. Their polarity would change to a positive charge and they'd become positrons but this would mean that there would have to be an inversion of the atom on another plane (anti-plane) or dimension. Where would this inversion exist?

The Standard Model is the list of what everything that exists is made of. It has been a beautiful tool into understanding the universe but it isn't complete. What we do know is that the proton and neutron are both made up of quarks. The first quarks were named the up and the down quark but four more quarks have since been discovered. The electron is classified as a lepton, which includes five other particles, including the neutrino. The leptons and the quarks are categorized as fermions, which includes particles related to matter. The rest of the model is reserved for physical forces, which are needed to hold everything together. The force particles are categorized as bosons. We have discovered where forces such as electromagnetism come from but gravity has been elusive.

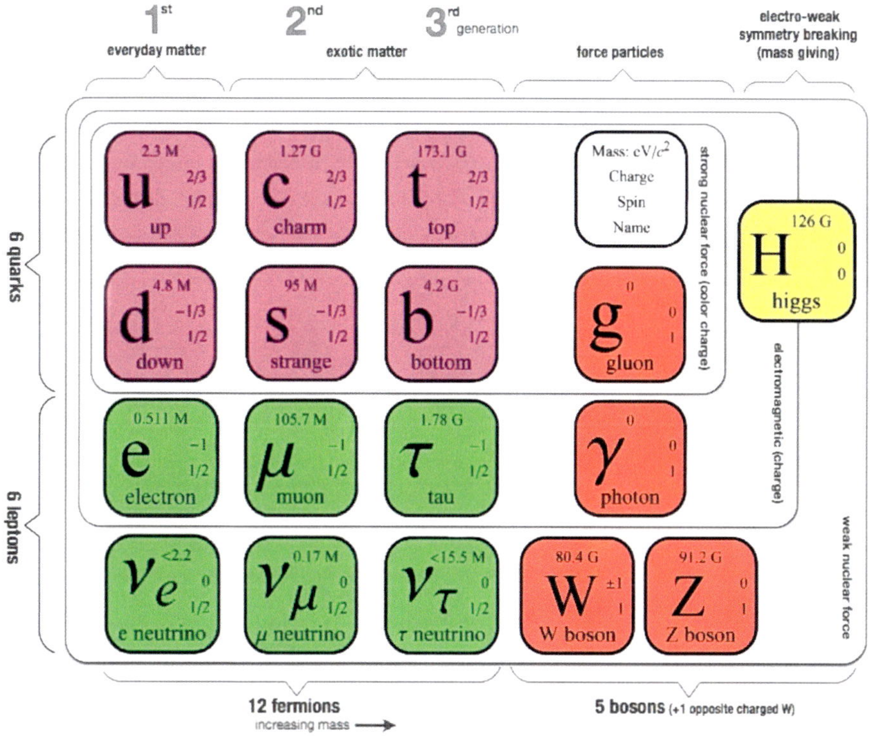

Gravity comes out of something without putting anything in; it exists no matter what. When the infinitely small is studied gravity is still present. The effect of gravity is instantaneous, so in a sense it's faster than light. If the moon were to suddenly disappear we would feel the lack of its gravity at the very instant it disappeared but we would still see it in the sky because it takes 1.3 seconds for the light to travel from the moon to earth. I believe that gravity is a doorway to a deeper understanding of Nature.

Quarks always exist in groups of two or three but every proton and neutron are made up of three quarks. You can never isolate a quark in space, since the force that holds them together gets stronger as they are pulled apart. The level of energy that is required to pull two quarks apart is translated into the creation of another quark, so if two are pulled from each other this would in turn create two more making a total of four. This exchange is widely accepted as $E=mc^2$ - but where does the matter-energy exchange come from? Is it energy directly translated into mass or does the mass and energy exist simultaneously on different planes trading with one another?

Particles also have what's called spin. This is not a literal spin in the sense of the word but a magnetic spin. This spin is described as an intrinsic angular momentum which is rotation without actually spinning. This is a bit hard to understand but I will try to explain. For example, each quark has a spin of either up or down and this spin gives the quark its charge. An up spin gives the quark a positive charge of +2/3 and a down spin gives a quark a negative charge of -1/3. A proton has a positive charge of +1. That's because its three quarks are made up of two up quarks (+2/3 & +2/3) and one down quark (-1/3) which is a total charge of +1 (+2/3 +2/3 -1/3 = +1). A neutron has a neutral or zero charge so its quarks are up, down and down (+2/3 -1/3 -1/3 = 0). This charge is produced by a magnetic field and this field is the particle's spin.

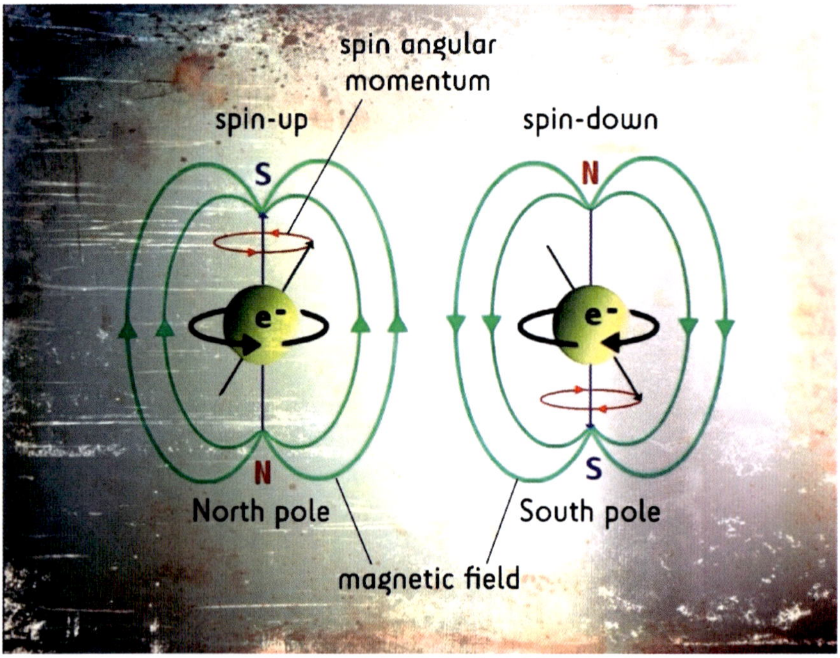

With all this talk about particles you may ask, "What then are strings?" You may have heard of String Theory. A string is a particle, which as I have already described particles can behave very strangely. Visualizing particles as vibrating strings helps us to understand some of this behavior.

A string is similar to a wave but its movement is much more peculiar. Strings can vibrate in the topologically equivalent of a circle (closed string) but they also seem to emit and terminate as a curve or a straight line (open string).

These modes of vibration equal different masses. String theorists have gone a step further to discover that everything starts to make sense mathematically if you account for ten dimensions. For as much as this helps it's still hard to imagine all of this taking place in the space a particle would occupy, instead of a nice round sphere that we've grown to be so comfortable with, but this is where the forces/bosons come into play.

The Higgs Boson is probably one of the greatest discoveries in physics. It was suspected to exist in the 1960s but wasn't officially discovered until March 14, 2013. The reason why the Higgs is so important is because it holds everything in the Standard Model together. You have to imagine the Higgs as a field that is part of the fabric of the universe. This fabric is made up of countless Higgs Bosons like the water molecules in an ocean. The more a particle (a.k.a. string) interacts with this field the greater the mass of the particle regardless of the particle's actual size. The top quark has 350,000 times the mass of an electron. If there was no Higgs they would both be close to zero in size and have no mass.

Various types of energy, force and particles have fields. These fields are all intertwined. They translate energy to each other through various types of stimulation. These energy exchanges can emit and absorb particles. This network of fields, in conjunction with the Higgs field, is the membrane onto which all matter, energy and life itself is projected.

When you put all of this together it starts to sound like science fiction:

- Electrons have the potential to exist in multiple places at once in either our plane or an inverted anti-plane with an opposite charge. They can also seamlessly teleport from one orbital shell to another.
- Everything in our universe is a particle and a wave vibrating at different frequencies.
- Gravity's reach can act on space faster than light and it affects everything around us but we have yet to find its source.
- Particles have intrinsic momentum, i.e. spin without rotation, so they seem have a momentum characteristic without moving. They vibrate as strings that behave wildly circling on themselves but also emitting out and terminating into the unknown. This whole dance operates in ten dimensions. That's six dimensions beyond what we can perceive.
- A membrane field not only holds everything together but is also the reason why everything has mass and is responsible for how matter changes.

Even with these miraculous discoveries you can see that there is still an underlying mystery here. My theory is an attempt to explain this mystery but first we must take a look at some of the current theories that are being explored.

SUPERSYMMETRY, THE MULTIVERSE & HOLOGRAM THEORY

Around when the Higgs Boson was discovered in 2013 it was a very exciting time in physics for a good reason. The further study of the Higgs and potential for more discoveries with the Large Hadron Collider were major stepping stones toward unlocking the mysteries of the universe.

There were two main groups of physicists that had their focus on what the mass-energy of Higgs was going to be when and if it was discovered. The first group was the Supersymmetry theorists. There's more than one version of Supersymmetry but for their theory to be a main contender their calculations showed that the mass-energy of the Higgs should be about 115 gigaelectronvolts (or GeV for short). The other group included the Multiverse theorists. There is more that one version of their theory as well but their calculations showed that the Higgs should be about 140 GeV. When the Higgs was discovered they determined that its mass-energy was 125 GeV. This of course left both groups scratching their heads. Some physicists have dedicated decades of their lives to their theory so you can understand how this would be a bit anticlimactic for some. It's like if you dedicated a great portion of your life obsessing/rooting for your favorite sports team and then they finally get to the big game, only to find that no one wins. It's not even a tie.

I personally love how evasive Nature is. If you haven't done the math yet 125 GeV is ironically almost in the middle of the two opposing teams. The other ironic part of this is that these numbers have a plus or minus factor to them too so it's very possible that the Higgs is exactly in the middle. The Higgs energy-mass of 125 GeV is also considered tentative for now so we'll see.

But what are the Supersymmetry and Multiverse theories about? I'm going to explain them as simply as possible but I will also focus on some key points that will be important later.

Supersymmetry sounds like what it is; a theory to try to create balance in the Standard Model so that the force equations are equal to the mass equations. One way to look at this is that there is a heavier cousin for every particle to account to the lack of mass and/or energy that we haven't discovered yet. Physicists love symmetry. It makes for beautiful math. Just look at $E=mc^2$. It's simple, symmetrical and beautiful - but Nature can also be unsymmetrical so should we look at mass and energy the same way as we look at the beauty of something as small as a tree or as infinite as the universe?

The Multiverse is a whole other animal altogether. It's the idea that we are one of a possibly infinite number of universes all with their own distinct characteristics that are bound to each other in a structure that isn't completely understood.

I'm going to aggravate some physicists and combine the Multiverse and String Theory, but to stress their individual importance in a constructive way, so I hope that no one gets their particle beams in a bunch. Both involve multiple sets of dimensions and universes, which are not the same thing, but in my theory they have an odd relationship so for the purpose of segue I'm putting them together.

I've already touched on String Theory but I want to reemphasize that String Theory calls for ten dimensions. This would be the four we all know and love which includes our three-dimensional space and time but there are six more. If we existed in two dimensions it would be difficult to conceptualize a third, because not only would we have evolved in two dimensions but our physical bodies/minds also exist in a two-dimensional universe. The common example is that of a water hose. From a distance it looks like a line but if you get up close it's not a line but a cylinder. On a quantum level the surface of the cylinder would be considered another dimension. This visual aid is just the tip of the iceberg. This is where the concept goes beyond our human perception just like the example of the two-dimensional man.

This is what physicists attempted to show what these particle dimensions might look like.

Can you see how these are fractal-like? A fractal pattern is a natural phenomenon that is also known as expanding or evolving symmetry. It's a repeated self-similarity; the whole looks like the part. Benoit Mandelbrot, an amazing mathematician, understood this and was able to apply his fractal math to everything in Nature by calculating how much natural "roughness" there was in any given thing. Fractals are in everything from your heartbeat to weather systems. It's part of every organic and non-organic thing in the universe. Everything from movie CGI to the antenna in your cell phone would not be possible if it wasn't for fractals. The fractal shape of your smart phone antenna has a direct relationship with Nature and is the reason why it has the ability to transmit multiple waves of information at multiple frequencies simultaneously. It's quite remarkable when you think about it.

Another radical idea called Hologram Theory is based on the concept that the entire universe can be viewed as a hologram. All of the information required to create this hologram can be stored on an event horizon as a jumble of data in two dimensions.

It was originally thought that when something falls into a black hole it's destroyed along with all the information that makes it what it is. This is not true. It is now believed that the information is stored as a two-dimensional fluctuation on the event horizon. The idea has been taken a step further to say that if there was a way to retrieve that information and put it back together that whatever fell into a black hole could be recreated down to the last particle. What if you could deconstruct and reconstruct everything in the universe in this manner? Once that was understood then physicists could study everything in the universe in a completely different way.

This two-dimensional information construct is scrambled, but if everything can be stored much more efficiently, then the universe can be studied from a completely different perspective. The hope is that this perspective will lead to a new piece of the puzzle that could have never been discovered otherwise. Unfortunately, I believe that this is more of an exercise for just that and not a way to spearhead into a deeper understanding of Nature. The information does not make the universe but it's the nature of the universe makes the information.

Even after exploring all of these ideas we're really not any closer to any solid answers to how the universe works.

The mass-energy of the Higgs turned out to be almost smack dab between the two estimates that would have supported Supersymmetry or the Multiverse theories. This means that both of their calculations, or the concepts they are using to derive them, are wrong. Maybe the reality is something in between.

Supersymmetry is trying to find balance in an unbalanced Standard Model by looking for particles that we can't find yet and may never find.

The Multiverse is overflowing with an infinite number of alternate parallel universes but even if this theory is somehow proven our further understanding of physics could come to a complete halt. It's not like we could ever study one of the countless number of these universes because we just don't have the tools. Even if we could somehow study one alternate universe we would never be able to see or understand the combined effect that the universes have on each other, if they have any effect at all.

Hologram Theory has its heart in the right place by expanding our tools of study and seeking backdoors to the mysteries of the universe. The part that deserves the bulk of our attention is the fact that the universe is not merely meandering over a scrambled bunch of ones and zeros, but represents a hologram.

HEAVY HOLOGRAM THEORY & THE PARTICLEVERSE

-AN INTRODUCTION-

What if I told you that there is a parallel universe that can answer all of the mysteries that plague the minds on physicists today? The concept of parallel universes always brings about ideas of science fiction, but the widely accepted theories I've outlined thus far undoubtedly breach the wildness of fantasy. I believe that there is math in everything but math is too static to be the starting point. The great minds of today try to "math" their way to the theory of everything because they lack the imagination of men such as Einstein and Tesla. Einstein predicted the existence of gravity waves. Our technology has only recently been able to confirm this but did we ever doubt their existence? Einstein's theories are so in line with Nature that they are undeniable to the open-minded. Our minds are limited; it's almost impossible for us to perceive higher dimensions, yet everything in the universe is linked by Nature. They're products of this beautiful universe and therefore our imagination is a natural connection to the answers we seek. We have to throw our minds into the chasm of the unthinkable to find the truth.

Imagine that there's a parallel universe that is so rich with energy that it's bursting at the seams. I believe universes are infinite so the "seam" I'm referring to is not the edge. It's a membrane, which is a very difficult concept to understand, but I will nevertheless attempt to describe it. You have to peel away your perception of your three-dimensional reality, which will fool you every time. Every particle, every smallest thing of a thing is blanketed by this membrane. The reason why this is hard for us to imagine is because we look at our reality like it plays out on a tabletop. This

membrane surrounds every particle like an ocean and the expanse of this ocean exists in a parallel universe, lying in the fold of the horizon that maps the fabric of our subatomic observable universe. This seems very intrusive to our tabletop reality. The reason why strings have broken the lead when describing particles is because physicists realized that particles weren't just points in space; they were vibrating waves of mass-energy occupying the area that you would define as a particle and, as I already explained, these "areas" have angular momentum with no physical spin. The spin and string behavior is a physical manifestation of the exchange between our universe and a parallel universe. The reason why we have drawn a line in the sand between the physics of the very small and the very large is because when you look at the world of the very small you can see the light from the projector that generates the hologram that we perceive as our three-dimensional reality. There is no difference between the physics of the large and the small; this is an error of our human perception. This parallel universe has so much mass-energy that it can affect our universe through the fabric of reality even in the vastness of space. I call the parallel universe that fuels this menagerie the Particleverse.

I imagine that the Particleverse is a very strange and beautiful place, but unfortunately it would be impossible to visit. I will give you two examples using our common perception to metaphorically communicate the nature of the Particleverse.

Imagine if you were suddenly transported to the Particleverse so that your entire body down the smallest particle appeared in the Particleverse at the same instant. You simply wouldn't appear; every particle would be instantaneously absorbed as energy waves, barely making a ripple. Now imagine if you were able to open a door and walk into the Particleverse. The moment your particles crossed the threshold they would instantaneously change into energy. The information needed to recreate you would still exist on the horizon and could possibly be retrieved but it would be scattered at the speed of light. The forces that bind particles and atoms together in our universe don't exist in the same way they do in the Particleverse - so neither would you.

If our eyes had the ability to see every wave it would be a spectacle, but this experience would be astronomically more intense in the Particleverse. The waves in the Particleverse are fueling every particle-string in our universe with a steady stream of energy that not only creates the mass of the particle-string but also gives it gravity. This stream of energy is like an infinitely long bow running along the string of a violin. This is what gives particles their spin. There are actually six streams of string energy in the Particleverse interacting with every particle in our universe at the same time. I will explain where this number comes from later but this is how the force of gravity is produced. The exchange of mass-energy between the two universes creates a vortex that bends space and creates the force gravity.

Every like particle in our universe exchanges information and energy through the Particleverse. In our universe this exchange can seem like it's faster then the speed of light. Everything in the Particleverse is still moving at the speed of light, but every unit or "area" of particle-string energy in this complex web of energy is like a mesh of gears working together. Imagine if there was a system of gears that surrounded the Earth's equator. If I turned

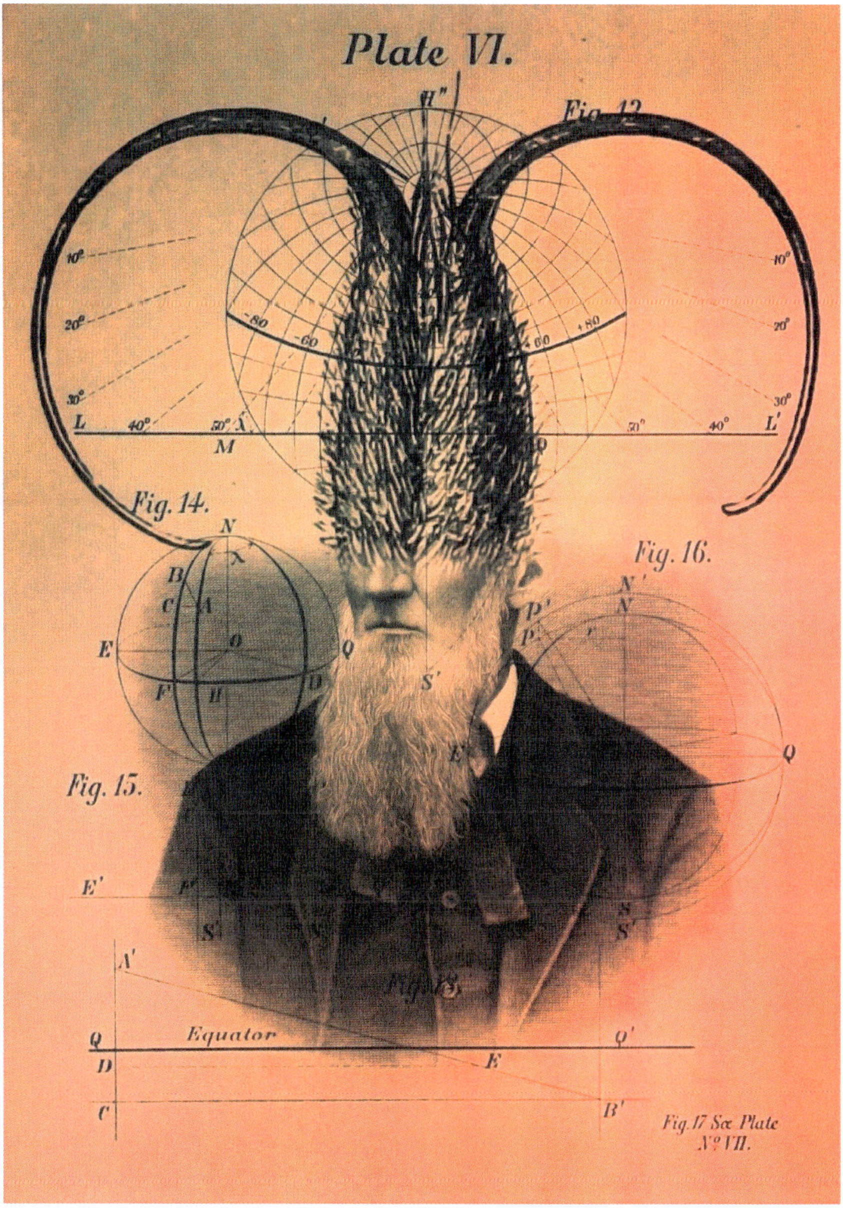

one in Brazil then a gear would instantaneously turn in Indonesia. In the Particleverse this gear system would also be traveling around the Earth like a train linked from end to end at the speed of light but in the Particleverse the path of this energy is not linear. The path of these energy highways is an organized chaos. Imagine every like particle has the same "color" or wavelength. The different colors of energy that fuel and create the colors of particles in our universe never mix in the Particleverse. These wavelengths of energy are a web spanning the six-dimensional space of the Particleverse. This is what explains the hidden communication of particles in quantum entanglement and why this communication seems like its traveling faster than the speed of light.

Quantum entanglement, or "spooky action at a distance" (as Einstein referred to it), is a physical phenomenon where the quantum systems of two or more particles become entangled through various types of interaction. Entangled particles are linked to each other no matter how far apart they are and this faster that light relationship has been proven to be appropriately correlated. This entanglement raises the possibility of transmitting information to the point of teleporting objects or even people in an instant. This would render fiber optics obsolete and could make the transporter of Star Trek a reality. In 2012, a team of European physicists successfully teleported photons a distance of 143 km (over 88 miles) using quantum entanglement.

The exchange of mass-energy between our universe and the Particleverse is the reason for everything. The nature of our universe is what provides the information for the existence of everything in our universe and the Particleverse provides means for information to exist. Our universe is a three-dimensional mass-energy hologram that is fueled by the Particleverse. I call this mass-energy hologram the Heavy Hologram. The fabric of our universe is a membrane in which the Particleverse transmits string-energy to particles. These particles have evolved their own material laws in our universe. I will explain this more later on but the most important thing to take from all of this is that you and everything in the universe are sharing particle-strings at the speed of light. Since living things are made up of like particles you are sharing particles with every living thing on Earth at this very moment.

Now I will explain how my theory satisfies the unanswered questions I've laid out thus far. I will also show how it connects the physics of the very small and the very large.

-LIGHT & TIME-

Before I delve into the marriage of supersymmetry and superstring theory I have to talk about light and time. Time is one of the four dimensions in our universe but how do we account for time in the Particleverse? Everything in the Particleverse is moving at the speed of light. If this were to happen in our universe then time would appear to stop, but there is a system in the Particleverse that is constantly connected with our universe. This relationship exists both in our time and at the speed of light in the Particleverse - so then what is time? Time is the moment required for the Heavy Hologram to manifest. It does not need to be constant. It just needs to be slower than the speed of light. That's why time dilation occurs as you approach the speed of light, which is actually a physical manifestation of the Particleverse. It's a product of the relationship between our universe and the Particleverse.

I've briefly explained what a quantum leap is. When energy is released from a quantum leap it manifests as light but I think that we take this conversion of energy for granted. According to classical physics an electron should emit electrometric radiation (photons/light) continuously because it's accelerating in a curved path around the atomic nucleus. The loss of energy would send the electron in to a death spiral toward the nucleus and in a very short time the atom would cease to exist. Obviously this can't be the case. Physicist Max Planck solved this important problem by demonstrating that energy on the subatomic level can only be transferred in small units called quanta. This unit became called Planck's Constant. Quanta are not divisible and quantum leaps are seamless transitions of electrons from one energy level to another. We know that electrons convert energy into light but we don't know how this is actually done. Planck's Constant is mathematically true, but why? This is all part of the balance between our universe and the Particleverse.

The electron is the rebel of the atom. All the other particles are married to another but the electron is whizzing around the nucleus in spherical orbits. The electron's job is to translate energy between the universe and the Particleverse. The Particleverse is laden with mass-energy and is moving at the speed of light. By the laws of physics and Nature, which have been created by this relationship, there are only a few types of particles that the Particleverse can share with our universe as a free agent. One of these is the photon. Every time an electron jumps orbits it passes through the membrane between our universe and the Particleverse. This is seamless but the electron that leaps from a higher orbit and appears in a lower orbit or vice versa may not be the same electron, even though they appear identical. Like I've already explained the Particleverse is exchanging particles with our universe constantly. We can observe this exchange in a quantum leap but this only can be observed as spin and string vibration in a quark. The Particleverse can accept any form of energy from our universe, but when additional energy is partnered with an electron as it crosses the membrane into the Particleverse a balance has to be maintained. The membrane is affected in a way that causes an equal amount of energy in the form of light (photons) to be released from the Particleverse into our universe.

Light and time have a special relationship. The closer you approach the speed of light the slower time passes. It's not just time but everything slows down. Time in the sense of seconds and minutes is irrelevant. As long as you have the quanta of Planck's Constant then the Heavy Hologram of our universe can exist. If time were to literally stop completely then the particle vibration and the electron orbits would freeze. The relationship with the Particleverse would be severed, the Heavy Hologram would turn off and our universe would cease to exist. Even if you were able to move at the speed of light time would still appear constant to you. Your relationship with the Particleverse would still work. Time would appear to stop for everything around you, but this is just your perception. Time really hasn't stopped so everything outside of the dilation is still business as usual. Now the question is, how can the same Particleverse have a relationship with you being inside of the dilation and with everything outside of the dilation? This is where things get confusing and our limited human perception becomes a problem.

The Particleverse, like superstring theory, requires a ten-dimensional space-time. One of these dimensions is time and three of these are the three spatial dimensions we can perceive. The remaining six dimensions account for the exchanges particle-strings in our universe have, as part of the Heavy Hologram, with the Particleverse. Every dimension is like a hand being laid on every particle. That's six hands in total with each connected by an arm. Each arm is a plane of an infinite number of particle-strings moving at the speed of light. This dance between our universe and the Particleverse has a bottleneck effect on the membrane between the two universes. This bottleneck demands a seamless pulse and this pulse is quanta, which is where Planck's Constant comes from. These six dimensions can also easily compensate for any time dilation in our universe. The six-dimensional system of energy has no basis in what we think of as time so it's theoretically impossible to predict its behavior with relation to time. The term I came up with for this relationship is Particle-Time.

When two particles collide in a particle collider under the same exact conditions it's still impossible to produce the same results. The reason why we can never get the same result is related to time: the Particleverse's interaction with our universe with relation to time is not being taken into account. These static fluctuations created by the energy relationship with the Particleverse may be impossible to measure because they are seamless, but if we could somehow predict the Particle-Time relationship without having to measure the fluctuations, then it could be possible to reproduce the same aftermath in a particle collision. Unfortunately we cannot observe the Particleverse. Even if we could "observe" it, our technology and our senses are useless because they are products of our universe.

After all this talk about time I have to take an unexpected turn in our journey to a theory of everything. I have to exclude time as a dimension and only focus on the remaining nine dimensions, particularly the moment or quantum. Our version of time is a product of our perception and now that I've established that time is nonlinear, I think it is necessary to exclude it. I'm not saying that it should be excluded from any and all equations that describe our universe; I'm just saying that this will be the last time I will essentially include it in my explanations. Now I can display a quantum cutaway view that will show an inside look of the nature of particles and their relationship with the Particleverse.

-9 DIMENSIONS-

I think I should take a moment to define what a dimension is. There are many misconceptions and ideas about the meaning of the word "dimension". Marriam-Webster.com defines it "**dimension** / noun / • The length, width, height, or depth of something • A measurement in one direction (such as the distance from the ceiling to the floor in a room) • The amount or number of things that something affects or influences • A part of something." If you believe that there could be an alien race made of energy living in another dimension then you may be thinking about dimensions in more of a science fiction way.

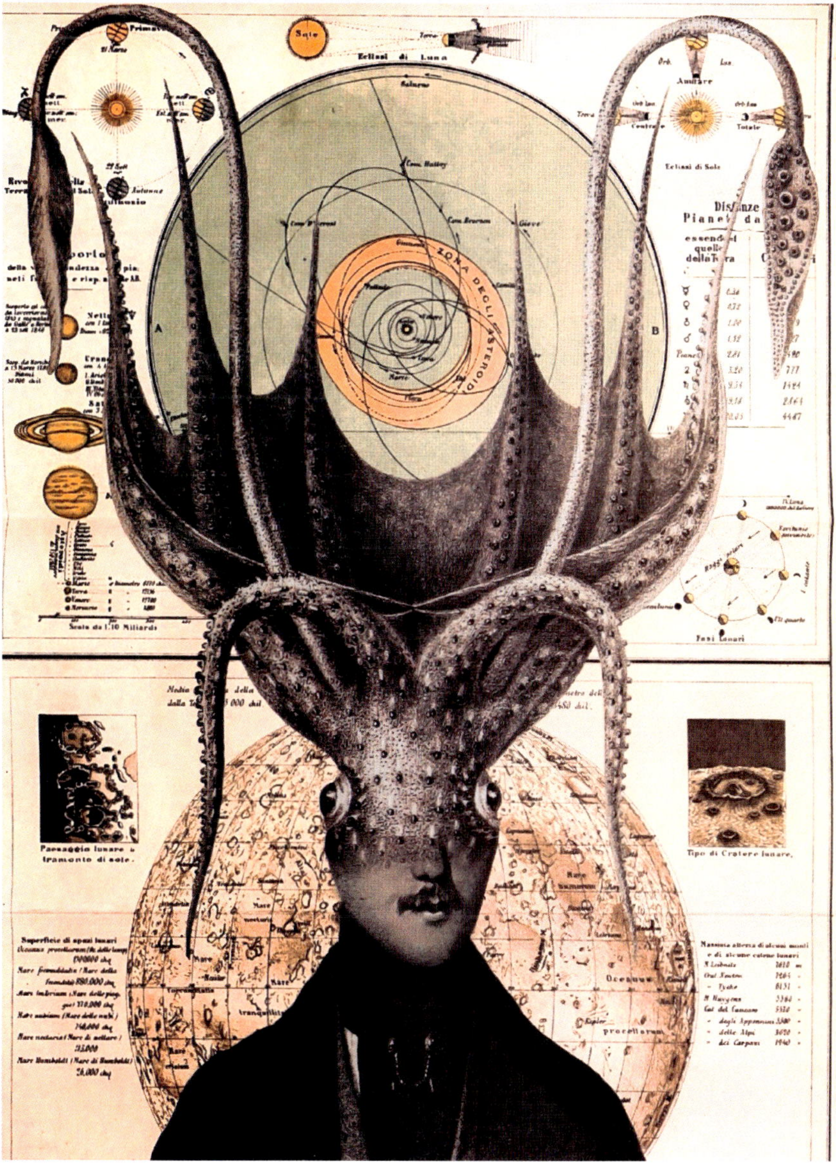

As everyone knows our universe is perceived in three-dimensional space. These three planes lay out the coordinates to our universe. The coordinates are laid in quanta and every coordinate has an energy value. Everything is connected because nothing is empty. The word "empty" should be scratched from the dictionary. The only use of the word that makes any sense is when you are using it to describe a feeling, which is perhaps another reason why we shouldn't use it. I'm joking of course. What would become of the popular idiom that relates a state of mind to a glass being half empty or half full? The point I'm trying to make is that there are infinitely large fields of energy that span our universe. I've already described one of these, the Higgs field.

The Higgs field is the outer layer of the fabric of our universe. I've also referred to this fabric as a membrane, the surface of which is the stage in which all creation is played out. It is known that interaction with the Higgs field is what gives particle-strings their mass-energy but this poses a question: where does this interaction come from and why do some particle-strings interact with the Higgs field more or less than others? In order for me to explain this I have to go back to the beginning: the Big Bang.

I'll be honest: there are a lot of things about the Big Bang theory that don't sit well with me. I think that there are some chapters that are missing from this story, but there is a wealth of supporting proof that it occurred in the way that physics has predicted and I'm not here to argue. My theories are not dependent on the finite chapters in the story following the Big Bang so I'm not overly concerned about any conflicts within the confines of this book, but I do want to share a hypothesis about the origin of the Big Bang. It could very well be a fantasy. Recently, convincing evidence emerged that could possibly prove the Big Bang never actually happened and the universe has always existed. I don't believe that this is correct either, but to completely ignore the possibility would be ignorant.

I will not be as bold in the specifics of the Big Bang as I am in the rest of this book about my convictions, because I don't believe that the Big Bang theorists have a right to be bold either. The problem is that the Big Bang happened in the past. We can look back light years in time by peering into a powerful microscope but that still only gives us clues into what really happened. What I will proclaim is that the Big Bang was a product of the Particleverse. The reason for the birth of our universe is still a mystery to me. I can hypothesize that the Particleverse was so laden with mass-energy that it demanded more dimensions to keep it in balance. The first crest or concentration of energy that was too much for the Particleverse to handle broke through as a singularity and simultaneously penciled the dimensions of our universe. These universes have been entangled in a dance ever since. There's a vertex between these nine dimensions (excluding time) that we cannot perceive because our senses have evolved in our 3D universe.

The Particleverse could just be another chapter in the evolution of our universe or the nature of the Particleverse could be to plant a mass-energy seed with the intention of reaping the energy back to itself. This is a far-fetched idea that I don't truly believe, but when I start talking about black holes, it seems more plausible.

Our universe first started with simpler elements like hydrogen. The nature of those elements interacting with each other under the force of gravity combined with the nucleosynthesis of stars created all the other elements. I'm simplifying things but trying to make it clear in an intentionally brief manner that gravity is greatly responsible for the creation of our universe. Gravity led to the formation of energy-mass clouds which eventually formed stars, solar systems and galaxies. The true beginning of my nine-dimensional-story begins with the birth of particles that had gravity. This happened just after the Big Bang. Before that there was nothing, so there had to be a stage in the expansion to allow the nature of energy to not only dictate its existence in our universe but to also spear the vectors of space. This stage was the Inflation period. Once this was all sorted the first particles were formed, the Heavy Hologram was switched on and space was a free-for-all. A long time passed before anything notable happened, but Rome wasn't built in a day. When Inflation was over the relationship between our universe and the Particleverse was fixed - OR - none of this happened and maybe there's an even older birth of everything.

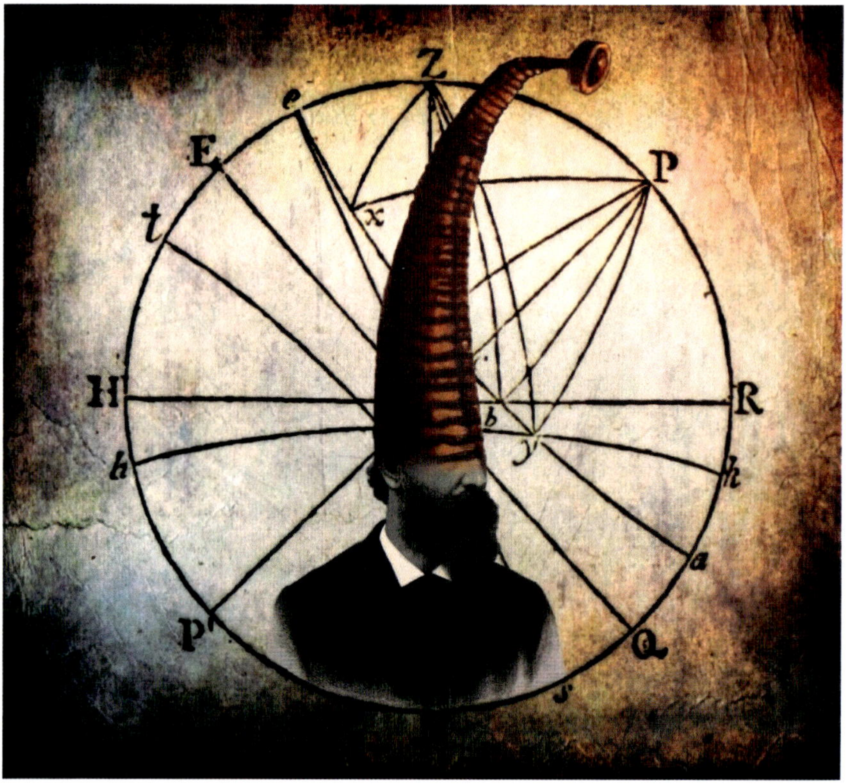

The dimensional vertex is infinitely linked to every particle-string. A reasonable balance is maintained but the energy of the Particleverse is never at rest. Our universe is always exchanging mass-energy with the Particleverse across a dimensional-vertex-horizon. The dimensions of the Particleverse are very powerful so therefore their influence has some side effects. There has to be a balance, but the Particleverse is so energy-laden with its six dimensions against our three that it creates a tug on our universe. You would at first think that a tug is unbalanced, but when you spread it across the vastness of space it starts to level out and even tips in the other direction. This "tug" is either directly or indirectly responsible for the creation of everything. It was the onset of the development of all the forces which led to the formation of atoms and governs the spherical orbits of electrons.

The exchange of the six-dimension mass-energy flow from the Particleverse over the dimensional-vertex-horizon through the surface of quanta-particle-string-bubble creates a fractal vortex. I know that sounds insane but please stay with me. This fractal vortex is responsible for the gravity force and also the intrinsic angular momentum of particle spin. It's the knitting needle to the 3D printer of creation. I showed this image earlier but I think it's important to look at it again under a different moniker. Here's a conceptual model of what the dimensional fractal vortex looks like.

This model is based on the ten-dimensional mathematical phenomena calculated by leading string theorists. This model is a fractal. Mass is directly related to energy because the level of energy that crosses the Higgs field through the vortex is what gives the particle its mass and the strength of the vortex is what gives the mass gravity. Therefore the level of interaction between the Particleverse and our universe is what dictates the force of gravity. This is also what grants gravity the ability to bend space. Everything is solid or not because of forces. Space itself is not empty and is also influenced by forces, which are malleable wave fields acting on energy-mass and the membrane between the two universes. This action can at times be very palatable to our perception - or it can fold completely out of our conceptual wheelhouse. The effects of gravity are instantaneous and unbroken so the gravity force can appear to be faster than light, but that's because it's always present.

The nature of the particle vortex is to draw mass-energy together to make more complex structures such as atoms and molecules. The fractal motion of the particle-string-vortex, combined with all the forces, creates a wake. This fractal wake echoes through atoms and molecules. The influence of this wake is very weak but its combined influence is the slight of hand in the fractal nature of our universe. It is the secret ingredient to the organization of matter. Everything from our DNA to solar flares are fractals. On an atomic level we don't consciously connect the energy behavior of a particle-string to the growth patterns of a tree but they are related. There are even recent studies that compare the shape and behavior of our brain cells to the expanding galaxies in the universe.

The Particleverse can also change mass into energy and vice versa like in a quantum leap, for example. Earlier I described what happens in any attempt to pull quarks apart: if you succeed, then the very energy used to accomplish this feat creates two more quarks instantaneously, making pairs out of the two that were separated. These bonds are governed by the Particleverse through the gluon. When energy is introduced into particles this starts to influence the relationships between the particles and the membrane, creating a gluon flux tube between the two quarks. There's a tipping point in which the tube becomes too long and too deep for the membrane to handle and must be filled to prevent a subatomic black hole. Therefore energy is preserved and the Particleverse restores order. We have exploited this relationship with nuclear power and nuclear weapons. Aside from the global impact, which would be enough, we also should respect this relationship as if it was a rain forest or a coral reef. We lose sight of this part of Nature because science doesn't have feelings and its beauty isn't always staring us in the face.

Before I move on to the next subchapter I think it's important that I touch on the numerological connotations relating to these nine dimensions before someone else does. Theoretical physics can sometimes get a bad rap based on the fact that the validity of some these theories can be next to impossible to prove, so the last thing I want in the aftermath of this book is to invite scores of numerologists and mystics to misemploy my theories. I do believe that there is some power in numbers but not in a mystical sense.

Aside from the obvious religious undertones, which we should completely ignore, there are three major types of numerology. These are the Chaldean, Kabbalah (or Qabala) and Western (or Pythagorean). The Kabbalah and the Western both use a nine number system. The Chaldean uses an eight number system because the nine is held to be sacred. I am also a huge Beatles fan but let's not get too tied up in White Album obscurities. Even though it would be fun to romanticize this ideology into my theory there really isn't any connection. It's just numbers against numbers.

There's one other thing that I don't want my theory to get confused with and that's vortex-based mathematics. I talk about a vortex and so do they. I use nine dimensions and the number nine is an integral part of their concepts but our ideas are so different that I wouldn't even label these as similarities. Their math only exists in our universe. They also make theological statements when they refer to their math models like, "Most great name of God," "High faith" and "Symbol of enlightenment." I don't think any individual has the aptness to label a math model in this way. There is one aspect of their math that I do find interesting and that is their application of digital root numbers but it was Einstein that said, "I do not believe in a personal God and I have never denied this but have expressed it clearly. If something is in me which can be called religious then it is the unbounded admiration for the structure of the world so far as our science can reveal it."

-A HOLOGRAPHIC SYMMETRICAL MULTIVERSE-

As I said earlier, when the Higgs boson was discovered they determined that its mass-energy was about 125 GeV. This mass-energy was the mid-point between the mathematical predictions of the Multiverse theorists and the Supersymmetry theorists. I've had my theories in my head since my late teens but when I heard this result I knew it was time for me to write this book. If the Multiverse wasn't the answer and Supersymmetry wasn't the answer then it must be something in between and I already knew what that was: the Particleverse.

I've already described the six dimensions as six hands being laid on every particle. Each of these six streams of mass-energy has a fluidic entry point into our universe across the vertex horizon. This energy fuels the Heavy Hologram at the speed of light and the forces hold it together. Aside from the mass-energy that is present during any given quantum there are also six partners waiting their "turn" surrounded by an infinite number of like particle-strings in the Particleverse.

Supersymmetry is in the search for the heavy or inverted cousins of the standard model in order to create mathematical symmetry. These cousins are there but not in the way the Supersymmetry theorists imagined. There are not just one but six cousins for every particle to make up for the lack of mass-energy. These six are right on the threshold of our universe and they have a significant influence on the standard model.

Every fundamental particle has this relationship with the exception of the bosons and the neutrino. These are direct manifestations of the Particleverse but the bosons and the neutrino manifest themselves in our universe in very different ways.

The neutrino is a very important result of the universe-Particleverse relationship and is probably the most peculiar. It's a revolving door that is generated by the Particleverse to balance the leaks between the two universes. Neutrinos can pass through anything because they are almost completely unaffected by mass. They themselves are also nearly massless and they travel very close to the speed of light. A neutrino can travel from the core of the sun to the surface in about 2 seconds while it takes a photon thousands of years to make the same journey. This is what makes the neutrino the perfect equalizer. Subatomic exchanges with the Particleverse can be slightly messy. If these small "messes" are not balanced then it could spark a chain reaction that could tear the fabric of the universe. Thankfully this balance is invariant and the natural result is the neutrino. The two bosons that mediate the neutrino are the W and Z bosons. They are responsible for the weak nuclear force and this force is responsible for the radioactive decay of subatomic particles. It's a fact that the neutrino oscillates its identity as it travels through the universe. It cycles between an electron-neutrino, a muon-neutrino and a tau-neutrino. This oscillation is like a lighthouse shining energy from the Particleverse. This also gives the particle greater adaptability. You can flood our universe with neutrinos without causing any harm and there's very little disturbance.

The massless bosons are the photon (carrier of electromagnetism) and the gluon (carrier of the strong nuclear force). The photon has the characteristics of a wave and a particle. You can think of this similarly as a .wav file on your computer, which is a digital representation of an analog sound wave. Your computer has to break up this sound wave into binary bits. Our ear cannot detect the gaps in the wave but they are there. In light these gaps are the gaps between photons, which are what causes wave-particle duality. Particles typically move in a straight line but photons move in a wave pattern. As I've already explained, photons are created by the interaction between the electron and the Particleverse. The electron is like a water sprinkler whipping photons into our universe, married into electromagnetic radiation. This radiation is both visible and non-visible (i.e. gamma rays, X-rays, microwaves and radio waves). Our eyes have evolved to see the light spectrum but whether this radiation is visible or not, the specific class of this radiation frequency is always divisible by three. If you look at the light spectrum specifically you see the three primary colors that make the three secondary colors. These frequencies are direct manifestations of the six dimensions of the Particleverse. These dimensions are distinct but we have to imagine that this vortex relationship across the vertex horizon is fluidic and complex. This interaction is like an artist's palette filling the entire spectrum with variations of color. The non-visible frequencies are just an extension of the same radiation span. I think one of the most amazing realizations of this dimensional interaction is that when you look at the sun you are seeing the Particleverse shine at you.

The gluon is a virtual particle that only manifests itself when quarks interact with each other. It's literally the "glue" that holds quarks together. They don't exist on their own but they have a color charge which means that gluons are virtually linked to an infinite number of other gluons that we can't observe, i.e. the gluon field. Like I've said before, nothing is empty and the gluon field is another field that can exhibit itself out of emptiness. This maybe a bit contradictory on my part, because I did note before that protons are made up of three quarks, but they can actually have five quarks or even more as long as there's an odd number and the additional quarks' charges cancel each other out. These additional quarks make fleeting appearances in the proton. The gluon field holds all of these quarks together with "force-troughs" called flux tubes. If you try to pull two quarks apart the flux tube grows between the quarks, but it's not stretched – the force tethering the two quarks together remains consistent. Eventually so much energy is introduced into the separation that it generates two more quarks, creating two new quark pairs.

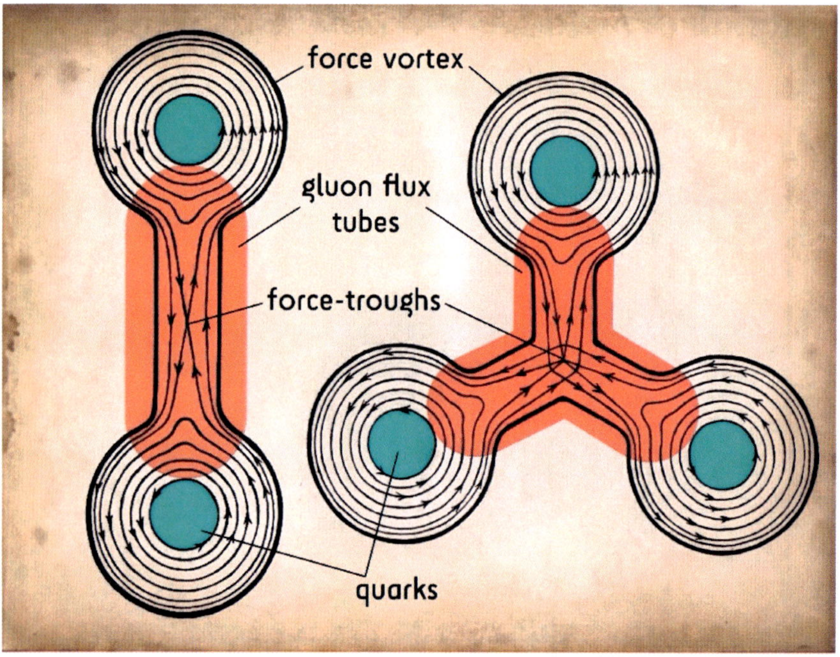

The Higgs field is what gives the subatomic particles their mass, but this only accounts for about 1% of all the mass you and I have. The gluon field is responsible the force-energy that makes up for the remaining mass-energy in all the objects in our universe. This is why $E=mc^2$ is so powerful. Mass and energy are synonymous and that's why there is so much energy in a small amount of mass. When these gluon force flux tubes are generated by quark interaction it creates a trench, which is like a subatomic vacuum that suppresses the gluon field. The quarks metaphorically surf on the edges of these suppressions and this is how they are held together. The gluon field is a deeper layer to the universe-Particleverse membrane. If you could look down into the trench of a flux tube you would be peering into a Particleverse abyss, but there are even deeper layers that keep the fabric together, so balance is still maintained but this abyss breaches our three-dimensional universe.

We are not able to see past the vacuum and I don't foresee us ever being able to understand or study these deeper layers, for these are the ones that exist in the Particleverse. Imagine an extremely choppy six-dimensional sea that exists on the edge of the Particleverse horizon. Wherever there is mass in our universe this sea becomes nine-dimensional. The force bridges the membrane to cradle the quarks together. The troughs of these waves only exist in the Particleverse and only the very tips of the crests exist in our universe in the presence of quarks. The waves of this ocean are like none that we've ever seen because they span nine dimensions and we only can model three of the dimensions of the gluon field, which is literally just the tip of the iceberg. When the gluon field is pulled too far and the trough becomes too long the Particleverse emits more quarks to preserve the balance. Bosons seem to disobey the Law of Conservation because they appear to throw subatomic "lassos" to feel out other lassos being thrown by other particles to establish force relationships, therefore creating something out of nothing. I believe that the universe-Particleverse relationship has to be perfect and could never allow this kind of anomaly. My question is: Why do we have to say that there is a betrayal of the conservation of energy when we are currently seeking a lack of energy-mass in the universe? The "choppy-nine-dimensional-oceans" are just one of many membrane layers that have considerable influence on our universe. These "lassos" are one part the fractions-of-a-percent that fills the lost mass-energy of the universe. My sub-chapter on dark energy and dark matter will shed more light on this conundrum.

Supersymmetry theorists want to unify the electromagnetic, the strong/weak nuclear force and even gravity. They've found that when energy is introduced the electrometric, the strong and the weak forces almost converge at the same point - but not quite. It's close but yet too far to show any serious correlation.

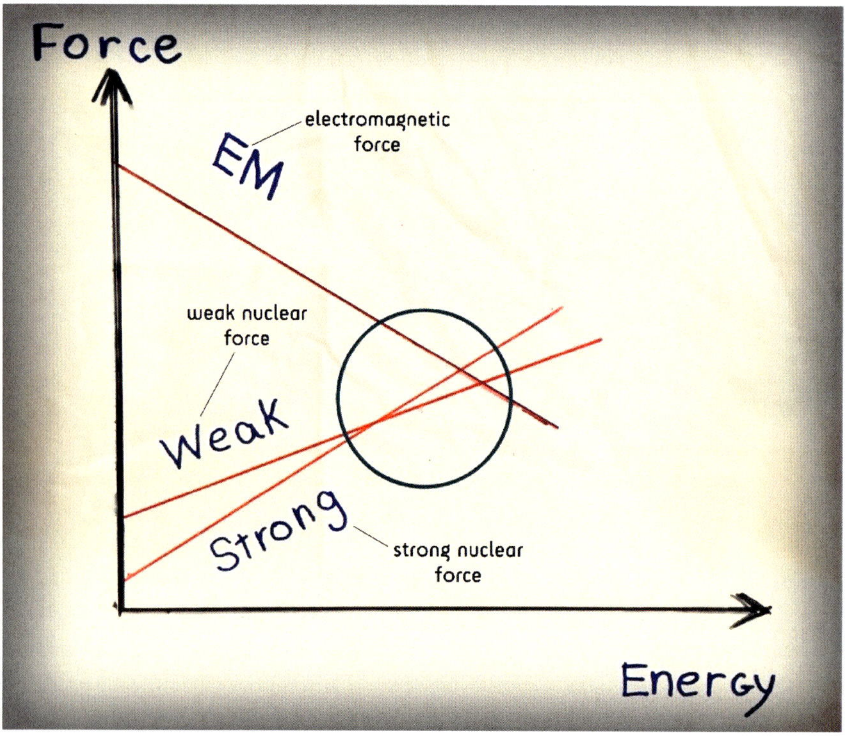

There is a correlation in that all of these forces are products of the Particleverse but the question is why then can we not observe this relationship? The answer is simple but solving it is very complicated. These forces are being bent across a vertex horizon which creates serious experiment control issues. You could say that the forces have a common denominator but they have specific jobs in Nature and these positions are layered by relationships that span over nine dimensions. We cannot simply put the forces on a two-dimensional graph and expect to solve the riddles of the universe. Even if we used a three dimensional graph it still wouldn't technically work, although it'd be closer.

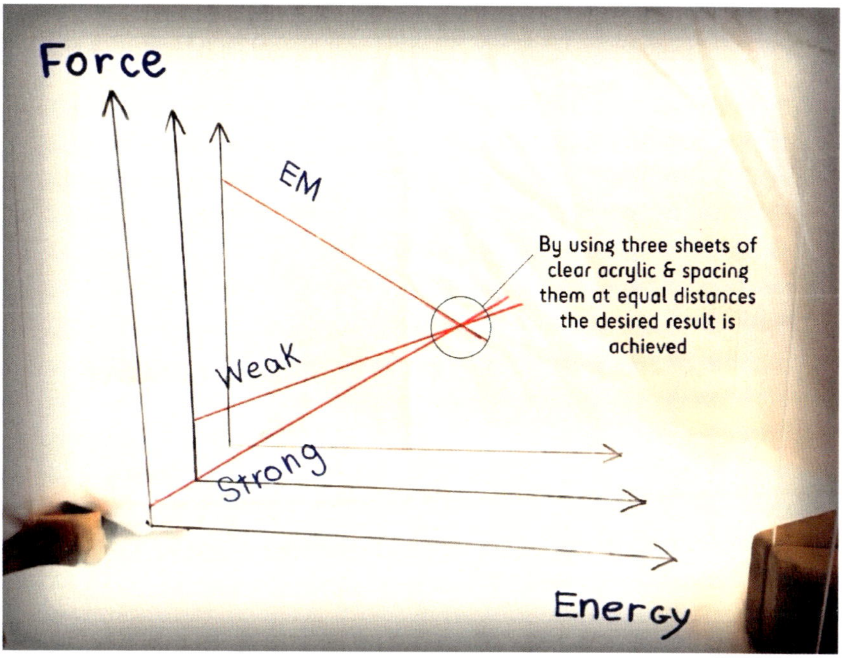

A personal issue I have with these graphs is that the "strength" levels have been sterilized to give a purer result. Nature cannot be simplified! My point is that we don't have the math or the comprehension to decipher this on paper as of yet.

The other part of symmetry is the idea that any particle can potentially become any other type of particle. Therefore a quark and an electron are essentially made of the same stuff. Its relationship with the Higgs field is what gives a particle its identity. This is somewhat true. I believe that the relationship between the universe and the Particleverse is much more complex. Energy is everything but the aggregate of the universe has been tightly woven by the six dimensions of the Particleverse and this weave has been sculpted buy the hand of Nature.

Nature is a very important word. I use it a lot in this book, but what do I actually mean by it? Nature is the root of the universe. It is the ever-evolving proverbial Ping-Pong match between the two universes. The common denominator of "Nature," or should I say the infinite exchanges of energy, have evolved into living super-clusters, living galaxies, living solar systems, living planets, living organisms and living subatomic particles. Everything has its own version of a breath. That breath consists of nine-dimensional fractal echoes.

In living organisms these energy exchanges have laterally evolved. We need to consume to keep ourselves alive. We all need to understand the idea that everything was connected and at some point sustenance became separate. Food demanded its own identity because the complexity of our evolution has to be mirrored by everything around us and that is the same evolutionary pattern for all life. It used to be a simple exchange of energy but those energy exchanges have evolved into you and I – as well as the consumption and reproduction of all living things. What makes us who we are as individuals is also a product of this evolution. All living organisms are connected because we all evolved from the same basic living material.

Our relationship with each other and our environment is what has evolved the energy that some would call the soul. Our capability for intelligence was obviously very important but when we found it necessary to have long-lasting relationships and to live in groups – that was the start to our modern sense of self. The development of relationships over generations of time is what gave color to social interaction and sculpted personalities. This evolution of self is part of the evolution of Nature, the evolution of our species and the evolution of energy. All living things, regardless of complexity, are still tethered together – just like they were in the beginning. Even though we have separated and evolved into the many wonders of life we've never separated from the Particleverse. It touches everything, fueling and exchanging information with the Heavy Hologram that makes up our universe.

The current Hologram Theory is great but it's being used for the wrong purpose. The senseless act of trying to deconstruct our universe by imagining that you can see it with 2D glasses across an event horizon is ultimately useless. You will never see Nature unless you use Nature to see it. Even if the concept is completely outside the scope of our senses and our understanding we still have to lean on a leg of our natural understanding to find the answers. Shuffling numbers in a raffle ticket drum will always give you an outcome but you'll never win every time unless you know how it works.

Touch your face right now. You are a composition of energy-mass and the nature of this structure has evolved all of your senses, so you're not conscious of it, but what you are feeling are energy-mass waves being held together by forces. This goes for everything around you as well. This is a complex construct of fields that interchanges energy, dictates mass and is governed by forces. On an atomic level every quantum fold is being bombarded and fueled by the six dimensions of the Particleverse. Your eyes have evolved to see the shine of the Particleverse emanate and reflect off the very universe it helped create.

Juli
Juni
Augustus
Mei
September
22 September
April
October
Maart
22 Maart
November
Februari
December
Januari

-DARK MATTER, DARK ENERGY & BLACK HOLES-

Dark matter and dark energy are just names for something we don't fully understand. We can observe the effect but we can't see the cause. The most acceptable idea I've heard for this is because dark matter and dark energy don't interact with the electromagnetic spectrum. This is true but it also has nothing to do with what dark matter and dark energy are. The reason why they don't interact with anything in our universe aside from their effect on space and gravity is because, in a way, they don't exist in our universe at all. There's a reason why we only can observe the effect of dark matter and dark energy. The Particleverse's interaction with matter in our universe creates the gravity force. That interaction is directly related to the behavior of dark matter and dark energy.

We only understand the mass-energy of approximately 4.9% of the universe. The rest is made up of dark matter and dark energy. Approximately 26.9% is dark matter and 68.3% is dark energy. If you compare how the gravity force affects our solar system to how it affects our galaxy we would conclude that the orbital spin of the outer edges of our galaxy would be slower than the center. This is not the case because the speed of the orbital spin is the same, so something must be tethering these celestial bodies together, keeping their orbits consistent and that is dark matter. Beyond the realm of galaxies the expansion of the universe is accelerating faster and faster with complete disregard for the gravity force. This accelerated expansion requires energy and this has been dubbed dark energy.

The supermassive black holes at the center of every galaxy are the heart of what holds galaxies together. Everyone thinks of black holes as universe vacuums and in a way they are, but they can also repel matter. The voracious appetite of a black hole is so powerful that when it feeds on large amounts of matter all at once it creates a super-hot spiraling disk. This disk is created by matter rubbing against itself due to the extreme changes of gravity and spin as it approaches the back hole. This disk is so hot that the pressure from the electromagnetic radiation (light) blows away the material that is beyond the black hole's reach, similar to a solar wind but much more powerful. This wind repels the mass outside of the disk into a safe zone. Some of the mass inside the disk survives by the way of star formation but the rest is consumed to the point where there is nothing left within the black hole's reach to feed on. The black hole stops growing and the galaxy becomes stable. This is the reason why these supermassive black holes don't eventually devour their galaxies. This border that is created during the birth of every galaxy prevents this from happening, but this tipping point is not entirely static. There can be periods of necessary feeding and fasting but the overall balance is still maintained.

Gravity is responsible for policing our universe. It creates, destroys, cleans, and preserves. This "preservation" has no consciousness or mercy but it does obey the laws of Nature which will, by default, tend to preserve life because it created it. The only time when we really run into a problem is when we try to disturb this relationship or when we abuse it selfishly. The Particleverse is a six-dimensional energy-mass sea that has peaks and troughs depending on how much matter it's interacting within in our universe and what proximity that matter has with each other. This exchange between all of the matter in our universe exchanging with the Particleverse is complicated but all matter in our universe is connected in this way. With more matter there is more gravity but when we take a look at our universe on a larger scale, the concentrations of matter swing dramatically and some very interesting things happen.

Dark matter and dark energy are two very different things but they are both a result of forces generated from the Particleverse. All matter is tethered through the Particleverse and the exchange of mass-energy creates the gravity force, but the webs of tethers create a compound effect on the forces in our universe. The tethers exists in the Particleverse but when there is a larger gathering of them pressed against the fabric of space it creates something like an inverted topographical-gravity map that influences all matter. I like to call it a gravitational chameleon field.

In the presence of a lot of matter it's a heavy field like in a galaxy. There's a healthy exchange of energy between all of the matter in a galaxy through the Particleverse even though it's so large so the average effect of this force is fairly consistent. This heavy field state even extends beyond the galaxy like a halo. This is what accounts for dark matter.

It's only in the extreme vastness of space where things change and this effect is dramatic. The field tips into an extremely "light" state. The exchange of energy between matter against the fabric of space becomes stretched from the lack of energy bleed into our universe which results in a higher frequency of particle anomalies. When this happens the field generates a negative pressure and acts like a cosmological constant driving the accelerated expansion of the universe. This is what accounts for dark energy. The chameleon field can have gradual or dramatic changes in pressure working with or against gravity. Here a simplified diagram to help visualize how the chameleon field works:

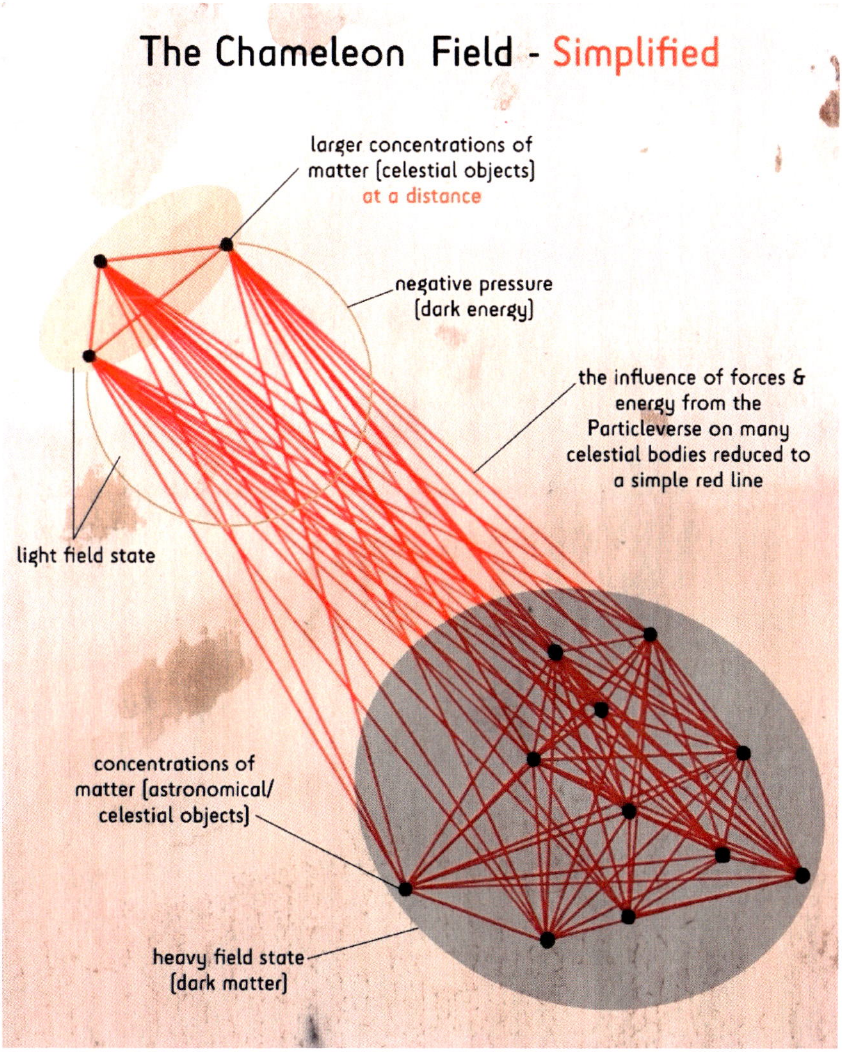

With the expansion of the universe there will be a time when we will not be able to see neighboring galaxies. For our species to be able to witness this we would have to move to a different galaxy before our sun engulfs us and Andromeda falls into the Milky Way, but it will happen. Even if darkness prevails beyond the edge of galaxies it will be more stable. A galactic collision like the catastrophic one we are expected to have in about four billon years would be almost nonexistent.

Earlier I used the analogy of echoes when referring to particles and their fractal connection to the construct of matter and all living things but the story doesn't end there. Hopefully by this point in this book you are

starting to understand that everything is "shoulder to shoulder." Nothing is empty and even if there was a scintilla of emptiness, the Particleverse is still there cradling the fabric of space. Even the smallest nudge has the potential to create a subatomic tsunami in an ocean of everything. These echoes evolve as they cross matter and space. The evolution of matter as these echoes travel from the world of the small to the large is characteristically different. Even though we can draw similarities they can't be simple mirror images because the layering of the influence of matter demands it to be distinct. An atom resembles a solar system and a solar system resembles a galaxy but they have evolved to behave very differently as a result of the compounded influence of matter. In the infancy of solar systems, like our own for example, the state of matter was more turbulent. There were more celestial collisions. Planets and moons were being bombarded with meteors and asteroids. Now the relationship within our solar system is more stable. The expansion of galaxies is just the next chapter of this stability.

There is also a connection between the supermassive black hole cut-off and dark energy. The force from the electrometric radiation that propels matter out of the danger zone doesn't stop there. It travels thought the galaxy field in all directions like a giant ripple and extends out into space. This aids in the expansion of the universe and is another way forces can naturally preserve matter.

We used to say that one section of space is similar to every other section we can observe. There's a lot of variation but we generally expect to see the same things. This isn't the case. Astronomers are finding these giant mega-clusters of galaxies that are a third of the size of our observable universe. I say that this is proof of the gravitational effect of the Particleverse chameleon-field. Even with the accelerated expansion of the universe there is still energy being exchanged between matter within mega-clusters spanning over a hundred million light years.

A black hole's gravity is so strong that nothing escapes - not even light. Black holes are a runaway effect of the vortex energy-mass exchange between our universe and the Particleverse. At this point in the book you may have already come to this conclusion yourself. Everything that black holes consume is absorbed back into the Particleverse. The information from everything being absorbed is preserved but scattered into a six-dimensional mass-energy sea. If black holes didn't exist then these accumulations of mass would just keep growing into giant celestial bodies that would engulf our observable universe. The increase of gravity approaching a black hole is so steep that in creates a barrier where matter is preserved and the excess of matter is devoured by the Particleverse. If the gradual accumulation of matter, like in the formation of planets and galaxies, were allowed to persist then it would never stop.

DETERMINISM

What is determinism and why am I writing about this in this book? Determinism is the idea that everything in the universe, including everyone's actions, is predetermined by a complex string of causes and effects. I feel Determinists may believe that my theory supports their beliefs but that's not entirely true.

We can draw the conclusion that the universe is generally deterministic. Laws dictate how everything interacts with each other but on the other hand we cannot predict when an earthquake will happen – at least not yet. The universal theory of causation basically states that everything that happens has a cause and effect. This includes philological determinism. Rational beings make choices based on how they feel. Feelings and emotions can be part of a complex part of a causal chain. This chain is a combination of influences that include society, upbringing, culture and environment.

Some determinists go as far to say that we can possibly eliminate moral anger by selling the idea that free will doesn't exist by focusing more on the causes of our actions and less on the punitive repercussions. This isn't a bad idea on some levels especially when you look at how class conflict affects society today but this idea creates major issues when you apply it to the world as a whole. Determinists may think that my theory may bring us a step closer to explaining away this randomness but I argue that even though it gives us the source, it is still random.

At first glance there are two ways to look at determinism with relation to my theory:

1. We are even more engraved in our destiny because our actions are bound by not one but two universes and the Particleverse accounts for all the quantum randomness. We are destined to be everything we will be even before we were born. Free will does not exist.

2. The Particleverse is completely chaotic and only manifests itself in a predictable way when it interacts with matter in our universe but the quantum strings of information and energy interacting within matter through the Particleverse cannot entirely be determined.

The reality is that it's actually somewhere in between these two. I find the subject of determinism very interesting but I don't admire all that could potentially come from it. We have to have an effect on causality otherwise everything becomes a slippery slope. Some "softer" choices that are

Pavlovian in a sense like whether I have a cake, ice cream or both (I like both) could very well be deterministic. I don't think a choice between murder and feeding the homeless fall into the same category.

My theory supports the idea of a collective consciousness so our thoughts and actions have the potential to affect causality two-fold but its effect is driven by laws of statistical averages – not by direct causality. It's a two-way street which is easy for us as a species to forget because it's our nature to be a bit self-centered. There are many causes upon causes and effects upon effects in play. A determinist would say that this is just a really complex string of antecedent conditions. It's not.

Imagine that our universe and the Particleverse are two mirrors facing each other. If you wave your hand between the two mirrors you would get this cool repeated effect going to infinity until it's beyond our visual horizon. Now imagine that these two mirrors are not flat – they have a roughness. Their roughness is different but no matter how rough they are they're still coupled by reflection. This roughness is an ever-changing fractal repeating itself into infinity and the opposing mirrors are still reflecting against each other into infinity. One of the mirrors represents our universe and it evolves in our time in three dimensions. The other mirror represents the Particleverse and it evolves in six dimensions in Particle-Time which is also independently dimensionally different than everything else in this analogy. Now instead of just your hand, imagine everything that exists waving between these two mirrors. Mark my words, someone will still say that this could be figured out – as insane as that sounds. I say it's impossible. When you look at all those different variables and apply those variables within the first simple concept of a repeated reflection into infinity – what you get is an infinite number of possibilities. Free will wins, determinism loses – game over.

As soon as our current level of consciousness and intelligence came to be it was a game-changer. Our future will never be static and we are not born into a predetermined sequence of events. Particle-Time will never be causal because its fluidity maintains the Particleverse mass-energy exchange with regard to any time dilation that occurs in our universe at any point in space. Until we have a thorough understanding of Particle-Time we can't have a serious discussion about this and even if we did, I feel very confident that we would come to the same conclusion.

We need to take ownership of our actions and be held responsible for our choices. If determinism is true then there is no hope for saving the earth because in a deterministic world we're just bacteria that's killing each other, consuming the earth and shedding our destructive waste in the process. We'll continue to consume and destroy until there's nothing left of earth but a cinder, because that is the deterministic path that we are on, but we do have the free will to change this. To say that that choice isn't there or doesn't exist is ignorant. We have an obligation not just to ourselves but to the world. Everything we put out into the world will eventually come back to us – good or bad.

THE COLLECTIVE CONSCIOUSNESS

What is collective consciousness? The idea of collective consciousness is that our species is able to share information, at least on a subconscious level, that manifests itself as repeated archetypes in human behavior. A common term used is a "hive mind," which in itself is an analogy. The reality of this relationship goes beyond our mind because the vessel that transmits this information is connected to every particle in our body.

We like containment. We feel comfortable within ourselves like a warm blanket, but part of our energy-mass exists on the horizon of the Particleverse. There is also a quantum string of binary information being exchanged with the Particleverse for everything at all times and we are part of this exchange. On the horizon of the Particleverse there is technically enough information to make two more three-dimensional versions of ourselves, which is a weird concept to comprehend. This is what I call our Quantum-Self. This version of us is a combination of the six additional dimensions of the Particleverse exchanging energy and information with our three-dimensional energy-mass at the speed of light. Since I'm on the topic of consciousness the most significant aspect of this is our brain.

Our brain has a hundred billion sensitive neurons. They can be easily affected by our environment so it's not hard to accept that they even be more affected by the six dimensions of the Particleverse. Our physical self and Quantum-Self combine to decide the ultimate aftermath of our neuroplasticity. Our brains act like a quantum chemical computer. If all the particles in our brain are constantly connected to the energy of the Particleverse then it's easy to conclude that changes in the brain transmit strings of binary information through this energy. Every quantum would be like a frame of a movie capturing an almost complete binary snapshot of the state of the brain. Our DNA combined with our chemical make up dictate the total spectrum of information that can influence us organically. The physical state of our bodies, including our DNA, are also part of this quantum snapshot. This is obviously a two-way street, like a breathing, chaotic mix of information between all life moving at the speed of light through the Particleverse. Life has the flexibility to change/evolve and the collective power of this infinite sea of binary information plays a heavy hand in the outcomes.

This binary particle influence can affect organs like our pituitary gland, which can in turn affect hormones that control our growth and development. Norepinephrine can be triggered by conflict because this is part of our survival instincts. Dopamine is created when our views are appreciated. There are also empathy neurons that allow individuals to simulate how someone else feels just by looking at them. These are a few of the important triggers that can have a profound influence on our emotions, our minds, our body and our Quantum-Self. Each neuron has a voltage which can change when ions flow in or out of cells. Once this has reached a critical level it will fire a chain reaction affecting other neurons where the process is repeated. This firing is married as waves and these waves oscillate in bands. These band waves are in turn mirrored in our Quantum-Self and transmitted as binary strings into the Particleverse.

In combination with our brains our bodies are also influenced. This particle influence has a hand in the mutations that lead to evolution. Genetic memory is incorporated into our genome from one generation to the next. This memory is influenced by our physical and emotional relationship with our environment but there is also a tremendous influence from the Particleverse. This is a natural way to communicate environmental changes quickly over a large populace to spark mutations that can potentially deal with global changes.

This is a beautiful relationship that all living things share. Life as we know it has evolved from the same origins. We have become almost devoid of our relationship with all living things but we are still connected. Our intelligence blinds us sometimes; we are so caught up in our individualism that we can't see what is really going on around us. Even though living things have evolved apart as distinct species we still share this communication because we are subatomically similar. It's very important to understand that energy has feeling. The energy that surrounds us and our Quantum-Selves is a beautiful chaos of emotion entwined with a living planet.

As we evolve there is repeated information that can be stored in our subconscious that can produce similarities in culture, behavior and how we interact with our environment. In the past there have been strangely similar developments in tool-making and architecture that have taken place over long distances without any possible contact. The similarities in pyramid-building are no doubt undeniable but there's no way that these civilizations had any knowledge of each other's existence.

We also share common ideas on what is appealing and interesting. Opinions are relative but there have been cultural phenomena in art and music that have bridged every sociological gap. At times everyone seemingly agrees that something is important and/or pleasing. I'm not just taking about the one-hit-wonder or someone's fifteen-minutes-of-fame; I'm speaking of global phenomena. Some of this is based on past cultural experiences but that can't explain it all. If the influence on collective consciousness is powerful enough it can win over a vast majority and the undecided start to succumb to the influence. In our distant past this was a way to help us evolve creatively and it also guided our decisions to ensure the preservation of our species but now, with all the information we're

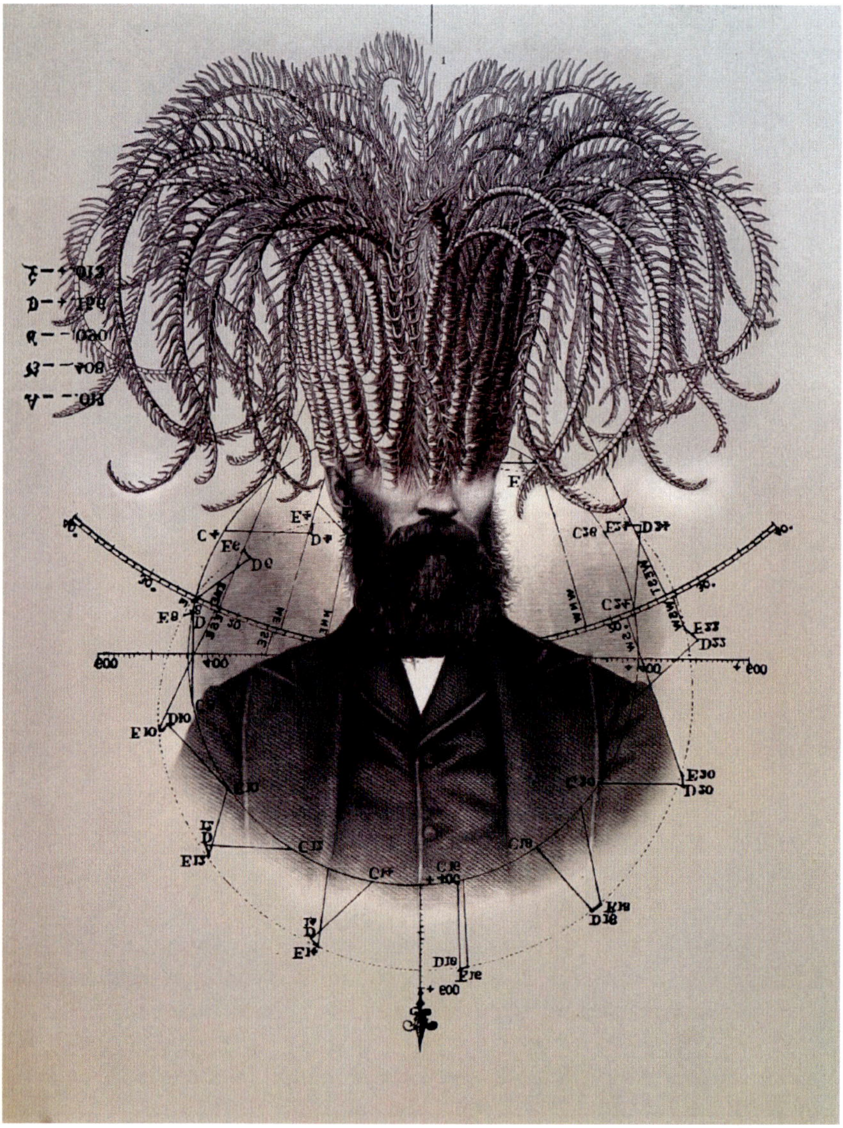

exposed to, it's become a lot more twisted. We have created a synthetic collective consciousness called the Internet. In a way, this competes with our natural ability to communicate organic experiences because we're flooded with too much stimulation, but this is child's play compared to what the current state of the world has done to this collective consciousness relationship.

Tragedy, stress and trauma in the extreme can resonate as markers in the chaotic binary information strings of the collective consciousness. In the chaos there are statistical peaks feeding back information that can influence our brains chemically and our bodies genetically. This information-string-radiation has unfortunately taken a sinister turn. During this explanation I've been using words like "us," "our" and "we" but I also want emphasize that this influence can bridge across different species as well.

We, as a species, have evolved with the ability to affect our environment in unnatural ways. Even though this influence only transmits small amounts of information, it's transmitting exponentially on global scale. If this does not change it will lead to dire consequences that are worse than any environmentalist can imagine and this is only one of many ways our choices affect our environment. This natural system has been in place long before we existed. We have a moral responsibility to try to maintain this balance to the best of our ability and we are doing the exact opposite. Money, resources, land and government won't matter because no one is safe.

Our environment is in jeopardy but this is an extension of a much deeper issue. We have become generally selfish as a species. We have created this pseudo-legacy-environment as a way to calculate our worth. We need to leave behind our positive emotional legacy as tribute to ourselves and the universe. Our whole global social dynamic needs to change. This is obviously a complicated problem but we need understand that by transitioning towards a more harmonic social dynamic we are still helping ourselves as individuals. The prospect of being selfless ultimately becomes a selfish act on a subatomic level. If we can embrace and share the understanding that all life is connected then maybe people would treat each other differently. Our species needs to have an awakening and see that we are all one.

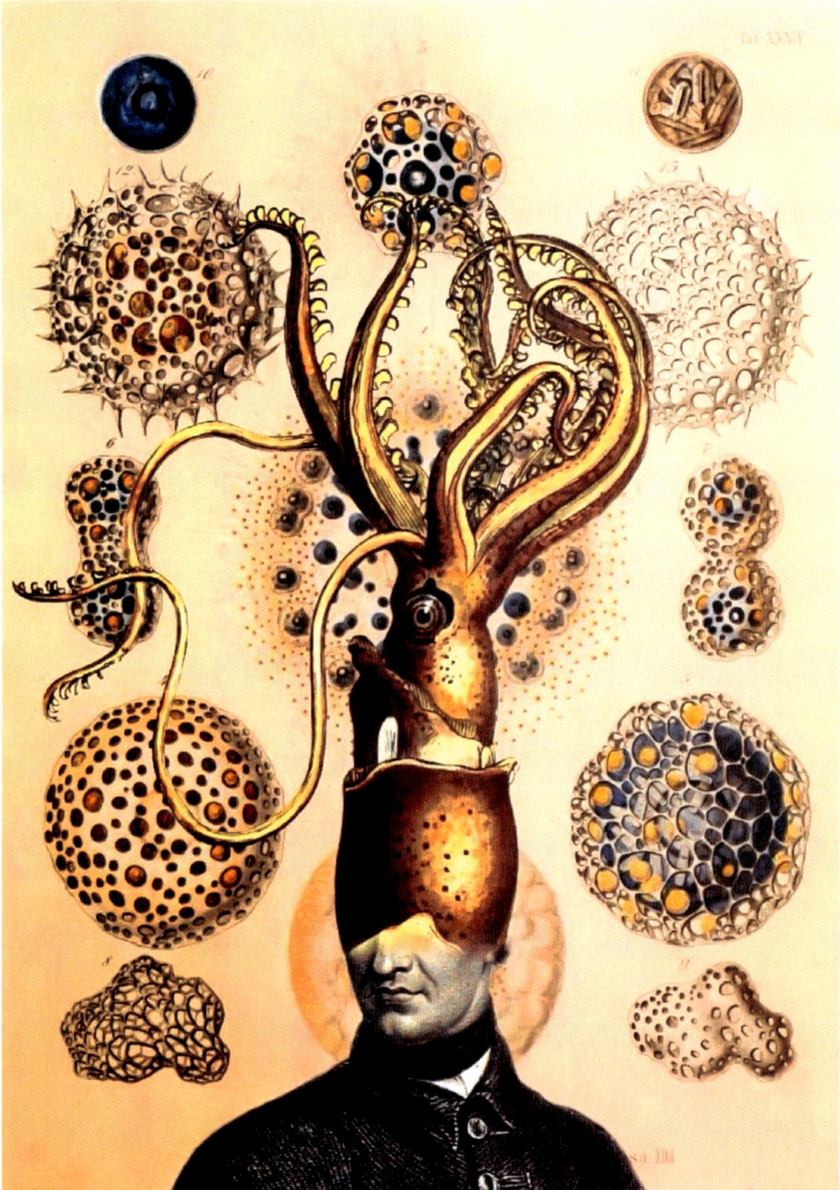

MORALITY

Everything we put into the universe comes back to us. Close proximity to the cause and effect of our actions is not necessary for this reciprocation but it can result in a greater impact. Everyone has the ability to influence their environment subatomically. This influence, under normal circumstances, is very minuscule so even if we were extremely efficient in detecting the effect of this influence it would still be virtually unobservable. It's our combined influence as a species that can have a significant impact on our environment - both negative and positive. The binary strings of information emanate from our Quantum-Selves combining in the Particleverse and then are instantly shared with our universe again. This information radiates out in all directions – following the energy path from one particle to the next – phasing between our universe and the Particleverse.

Every quantum snapshot of our physical state carries a tremendous amount of information and every snapshot is a frame to the movie that documents our life from the point of view of our Quantum-Self. This movie is about everything inside us all at once. It is difficult to perceive because we only have the ability to focus on a few things at the same time. Some of the information included in the picture comes from our thoughts, dreams, memories, emotions, physical health and DNA. A lot of this information is common so even when you take our combined influence into consideration it still has little to no effect on our environment. When the information is part of our unique experiences, thoughts and actions then it has true power. This portion of the information string is also by nature the most active. I could attempt to break down this information into categories and subcategories but that would just distract away from its significance. It's more important to understand that trends in the binary string spectrum develop overlapping concentrations in the overall mix of information. For example when you look at happiness and sorrow they

have the potential of creating peaks and valleys in the binary information mix that act like markers. If there is an extreme level of sorrow in the world then there would be a concentration of this "color" in the emotion portion of the information spectrum. That would have an impact on the combined living psyche. As I mentioned before this impact is not just limited to humans. Humans are emotionally complex and this complexity can vary greatly from on person to another so some of us can be more susceptible to these types of markers. On the physical side of the spectrum, the portion that includes health and DNA, these markers give our bodies a general awareness of our environment. The introduction of unnatural substances into Nature and our bodies on a mass scale can result in markers that have unpredictable repercussions. As you can probably imagine there is almost an infinite number of various combinations of this information that colors the spectrum of every quantum.

The binary information strings are rippling out from everything by following the path of the particle-mass energy exchange of the Particleverse. This leads to the conclusion that the intensity of the information's influence can vary depending on ones proximity to the source. There's a global mix that relays the general state of organic life and there is a proximity mix that transmits a more localized state. These layers work together in hand with Nature to assist and impede life for the benefit of the whole. In extreme cases this can create a Quantum-Self "microclimate" that is partially responsible for a variety of things ranging from riots, cults or the spark of a political movement. These microclimates can even have a negative effect on the heath of living things that are exposed to them depending on levels of stress and trauma that are fueling it. Types of environments like deadly battlefields and large tragedies like the ground site of 9/11 are good examples. These examples obviously had/have their own heath risks but exposure to these types of environments can also have a negative effect on us subatomically.

The beginnings of these proximity microclimates can start with only a few or even just two individuals spiraling out affecting a larger group. These types of actions are exceedingly dramatic. This is when our actions have enough power to affect as little as two living beings. The worse example of this is the act of killing. To make matters worse there are many killings taking place every day all over the world and we justify this with the illusion that there's a need to do so. I understand that this is a complicated topic so I'm sorry for simplifying and I don't mean to disrespect anyone.

What kind of binary information do you think our body transmits to the Particleverse when we die? On a more serious note - what do you think it translates when we die violently by the hand of another of our own species? Some animals kill within their own species as a matter of survival because they're naturally programmed to, but we choose kill. The consequence of this act is a proverbial red flag in the binary information string.

In the quantum world there is a version of karma that is very real. Everything we put out into the universe comes back to us. If an individual affects someone in an intense way the binary strings generated and radiated as a result of that action have the potential to physically and mentally affect the individual as well as all the surrounding life. The information that transmits through us can affect our heath and the heath of our unborn children. With every extreme action we're changing the angle of our life path of our Quantum-Self. Even a small change in course over time can result in giant deviations. We are actually helping ourselves by following the Golden Rule as well as following good moral and environmental standards. Unfortunately society makes this a difficult task. In some parts of the world this is practically impossible.

We are broken. What has happened is that we've made some bad choices and the ability to fix these choices is in the hands of our leaders. The cards that we're dealt as individuals restrict us from aligning with Nature. We have to remove ourselves from society, which isn't feasible for everyone to do at same time, or we have to revolt against the powers at be which ultimately feeds more negativity into the quantum binary string. The most practical way to get to a solution is for all the leaders of everything to recognize the dire need to solve the problem and to realize that by doing so they are solving their own. The leaders of countries, religions and even terrorist groups need to understand the importance their actions and recognize how it affects them and their people.

Energy has feeling. The modulation in the energy that is being exchanged between our universe and the Particleverse is flooded by the troubled binary emotional state of the world. The decisions of our leaders create conflict among themselves. This conflict trickles down to everyone else. Even under the best circumstances it can be difficult for them to choose the right path because they are also coupled to the decisions of other leaders that prevent them from making the right choices. What this has created is a perverted exponential growth of negative emotional states. Our minds' consumption of the media and the Internet, with the current condition of our society, compounds the assault on the state of our Quantum-Self. The environmental and emotional impact of our actions is partially responsible for every school shooting and every suicide bombing. Children are even more susceptible to this influence which is even more of a reason why we have to make changes.

The first thing we need to do is to be conscious of this influence and to second guess our own choices when they fall in line with a large populace. This is a difficult proposition to fill. We have to initially be more aware of our core values and try to avoid knee-jerk reactions/decisions based on emotion. This takes a degree of discipline that some of us may not possess. Leaders take advantage of this influence right now without really understanding why it works as well as it does. If we were more sentient of this effect then we would be able to make better decisions. The second goal would be to convince all the leaders that they can still pursue their interests as long as it doesn't dump garbage into the binary particle string. It's impossible to make this leap right now but we have to start initializing this idea as soon as possible otherwise all will be lost. If they all follow this principle then everyone, in theory, would be satisfied because greed would be abolished. It will probably be up to a younger generation to eventually make the final transition.

The leaders of the world need to understand that their control is only an illusion. They also need understand that if their decisions dump negativity into the quantum information string - even if they feel that they are helping their people by doing so – that in the end it will only perpetuate more negativity for their people and themselves. For whatever goal any leader is trying to accomplish – it's not going to matter unless they are conscious of the subatomic impact of their actions. "Do unto others as you would have them do unto you" is a quantum reality that no one can escape. The idea that a military campaign, consisting of something like a bomb strike, can support a greater good is ludicrous. These types of actions perpetuate an equal amount of destruction on our Quantum-Selves by either affecting our future choices or by subatomically tweaking our physical heath potentially for generations so if the leaders of today think that the eventual aftermath from these types of campaigns will lead to a utopian society for their people – they couldn't be more wrong. Any goal path constructed of stepping stones made from the loss of life will only lead to physical/emotional sickness and even more death. The upper hierarchy needs to make the right choices for themselves and for their people or this could all eventually lead to our extinction.

LIFE & DEATH

What happens to our physical consciousness when we die? I call this our "physical" consciousness because I'm purposely diverting away from religious connotations while focusing as much as feasibly possible on physics as we traverse something that is still considered phenomena. Life is the opposite of death therefore life ultimately defines death. There's of course the medical classification of death, such as brain death, but this is still part of philosophical debate. Women that have been technically classified as brain-dead have still been able to gestate a fetus and we've all heard of stories of people that have been revived after being clinically dead, which further begs the question: what is death? Dr. Duncan MacDougall was famed with his findings that the soul has weight. He weighed six patients that were in the process of dying and determined that there was a weight loss of 21 grams at the moment of death. This was rightfully discredited but logic leads us to believe that something must be lost. The answer is energy but the modulation in this energy is unlike any other.

When our lives start in utero there's an instant when our consciousness is formed. There's a development period before and after this moment that I like to define as a chain of connections. The chain leading up to this moment is a gathering of different elements of what some cultures define as a soul and the chain that follows is to further develop these connections, by adding layers that may not reveal themselves until later in life. As I explained earlier our brains are quantum chemical computers. The energy that fuels these computers not only powers them but also carries information. We know so much about so many things but we still know so little about our own brains. I believe that the reason why this is such a difficult hurdle for us is because we are not thinking about the brain in a complete way. We're so focused on the biology while ignoring the energy behind the biology. There are two things simultaneously changing in our brain at all times. The first is our brain anatomy, mostly chemical but

also physical and the second is the energy needed to accommodate these changes. Changes in brain anatomy are very complicated and very specific from one person to the next. Aside from the many chemical triggers that affect emotion and hormones, humans are also gaining and losing brain cells depending on their age and environment. Connections in the brain are made and broken depending on the need or lack thereof. Even though the results from these influences can drastically vary they still have to remain within the spectrum that was originally laid out before we were born. This is like our character DNA but is more energy-driven than biologically-driven. I believe that the fractal modulation of the energy that makes life possible is also responsible for forming the biological neuron fingerprint that dictates our individuality.

Imagine a complex fluid computer that exists in nine dimensions (ten if you include time). The integration of the complex chipset of this computer adapts, evolves and is ultimately laid out by the nine-dimensional fractal modulation of the energy that fuels it. The particles, atoms, molecules and eventually cells are guided within the barriers of the consciousness spectrum by this energy, therefore we are this energy. Then you might ask, "What is controlling the construction of these soul-programs?" The answer is… nothing. There are some character traits, skills and intelligence that you get from your biological parents but what essentially makes us who we are is a random construct that is built after we're conceived. The particles that make up our developing brain cells become linked to the Particleverse and the energy that feeds these connections is a scattered first-come-first-serve sea of constantly moving information contributed by everyone that has lived before us. This mix creates a progressive chain of recycled consciousnesses. The collective conscious and our experiences assist with the living evolution and death towards the final summation of this energy-information. The mental portion of our six-dimensional quantum-self becomes untethered from the consciousness of our universe and is absorbed into the Particleverse. During the course of our lives we contribute to this cycle by converting typical energy from the Particleverse into energy that can be used to fuel and give birth to more life. This is how an ever-growing population can be fulfilled if needed. Some may perceive this as a mild version of reincarnation but it's still just information. There is no one-to-one exchange of consciousness. If there was then there would be less development, but nothing in Nature is linear. We are all carriers of subconscious bits of knowledge and experiences of the thousands of generations that lived before us.

We are all one of the same energy. All energy is mass and mass is energy. It's miraculous how Nature has given the means for the existence of life. All of this was set in motion after the Big Bang. One step led to another till we found ourselves nestled in our corner of our galaxy. Most of us take this for granted. We're an overly complicated species that's too focused on our individual idealism. We have been given such a gift. We waste our energy on superficial sociological ideas and fail to see the whole picture. Everyone needs to stop what they're doing, pull their head out of whatever mindless trench they are stuck in, take a serious look around and wake up. We are entwined with a perversion of Nature that in every aspect

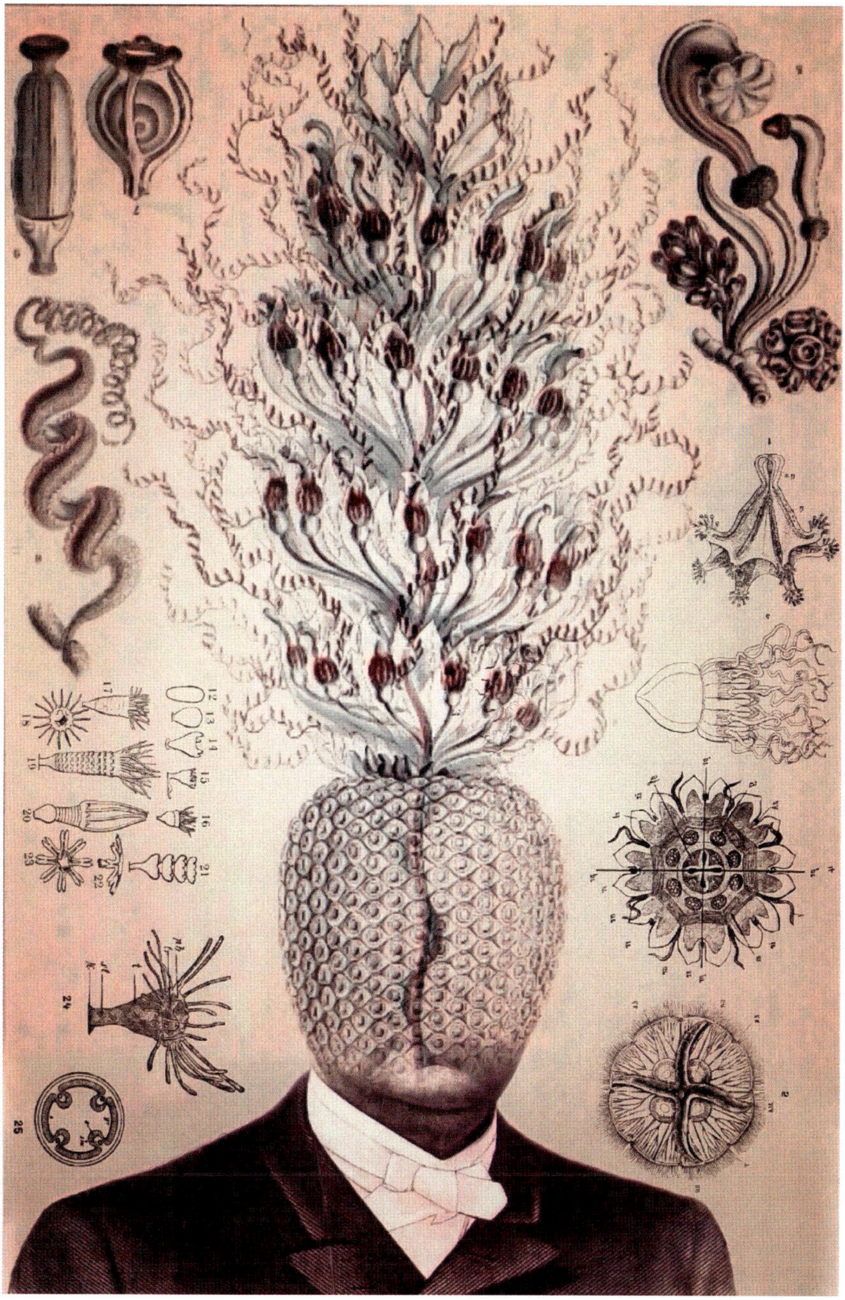

and on every level is completely out of control. Some of our natural connections to Nature have been twisted into situations that are harmful and self-defeating. We need to be more conscious of our relationship with the universe, with Nature and with each other. We also need to embrace and respect these relationships with intelligence and compassion. During the development of civilization we went astray. Instead of fighting the situation and each other we need to realize that it's not too late to get on the right path. If we stripped society away till we're left with just us, then we would see that we are actually in control and begin to grasp that we have the full capacity to solve all the world's problems. The human race needs a Renaissance of global proportions in order to survive the current state of the world. I want to be part of the Renaissance… What are you going to do?

ABOUT THE AUTHOR

Cash Mattock is most well known as a musician and digital collage artist. He was awarded a full scholarship to study engineering at R.I.T. and he also studied music at Nazareth College. Cash has had an obsession with theoretical physics since he was a teenager. He has been a guest on local, regional, and national radio, and he also has done numerous print interviews. His artwork has been featured in various digital and hardcopy publications. Cash has over a decade of consulting experience in the music industry involving management, A&R, P.R., marketing and legal. He has recorded and produced several albums, including one of his own, as the owner of CMI Records (Cash Mattock Inc.). Cash is currently working on his long awaited ground breaking second album that will be released in 2017. For more information please contact CMI at P.O. Box 149, East Rochester NY, 14445, or log on to CashMattockImagery.com and CMIRecords.com.

Made in the USA
Monee, IL
08 October 2024

67419109R00050